DESIGN DATA FOR RECTANGULAR BEAMS AND SLABS TO BS 8110:PART 1

A. H. Allen, MA, BSc, CEng, FICE, FIStructE

Spon Press
Taylor & Francis Group

LONDON AND NEW YORK

Viewpoint Publications

Books published in the Viewpoint Publications series deal with all practical aspects of concrete, concrete technology and allied subjects in relation to civil and structural engineering, building and architecture.

12.092 First published 1987
Reprinted 2004
by Spon Press,
2 Park Square, Milton Park, Abingdon, Oxon, OX14 4RN

Transferred to Digital Printing 2004

ISBN: 0 86310 029 5
© **Palladian Publications Limited**

CONTENTS

Introduction

In BS 8110:Part 1:1985, the British Standard for the structural use of concrete, two methods are given for finding the area of reinforcement required for a particular design moment at Ultimate Limit State when the section size is known. One is by using Design Charts, Part 3 of the Standard, which have been prepared from the rectangular parabolic stress block for the concrete in compression, and the bilinear stress strain curve for reinforcement. The other method is by applying formulae which rely on the simplified rectangular stress block for the concrete in compression. Although the first method is more accurate, the difference between the two methods is of little practical significance. The tables in this booklet have been prepared using the formulae and as an additional design aid, not as a replacement for the design charts.

Similar tables were prepared in accordance with CP110 and were found to be particularly useful where small resistance moments are required. It is sometimes easier to read the answers in tabulated form, where intervals are quite small, than it is to interpolate from charts.

As the majority of reinforcement used today is high yield, one grade of reinforcement, 460, has been used with three grades of concrete – C30, C35 and C40.

Derivation of values

Clause 3.4.4.4 of BS 8110 uses the term $K=M/bd^2f_{cu}$. In these tables the value of M/bd^2 is used for the different values of f_{cu}, and is preferred to the Code term as, not only can answers be compared directly with the design charts, the value of M/bd^2 is also used when checking deflection ratios.

(a) Sections reinforced in tension only (singly reinforced).

The two basic equations used are

$$A_s = \frac{M}{0.8f_y z}$$

and $z = d \left\{0.5 + \sqrt{\left(0.25 - \frac{1.5M}{1.34bd^2f_{cu}}\right)}\right\} \not> 0.95d$

Note – in BS 8110 the second term in the square root bracket has been simplified to $M/0.9bd^2f_{cu}$

If $R = M/bd^2$, $p = 100A_s/bd$, $f_y = 460$ N/mm^2

then

$$\frac{z}{d} = 0.5 + \sqrt{(0.25 - 1.5\,R/1.34f_{cu})}$$

and $p = 0.25\,R/\,(z/d)$

For a particular value of f_{cu} we can find values of z/d and p for differing values of R.

The limiting values of R, which will be called R', for singly reinforced sections are given by

$R' = 0.156\,f_{cu}$ with up to 10% redistribution

$R' = 0.132\,f_{cu}$ with 20% redistribution

$R' = 0.104\,f_{cu}$ with 30% redistribution

The values at which these occur for 20% and 30% redistribution are indicated on the tables.

(b) Sections reinforced in both tension and compression (doubly reinforced).

When the value of R exceeds the values of R' given above, depending on the grade of concrete and the amount of redistribution carried out, compression reinforcement is required as well as tension reinforcement.

If redistribution has been carried out it is not practical to present tabulated values for all possible percentages of redistribution. Neither is it practical to have several values of d'/d, the ratio of the depth of the compression reinforcement to the depth of the tension reinforcement.

The tables for doubly reinforced sections have therefore been prepared for the grades of concrete as for singly reinforced sections and for values of d'/d of 0.10, 0.15 and 0.20. Each table gives the percentage of compression reinforcement and tension reinforcement depending on the redistribution percentages of 10, 20 and 30. The starting point in a table is therefore the value of R', or the nearest value to one decimal place above, as given above. For example, using Grade 30 concrete and 10% redistribution the table starts at a value of $\dfrac{M}{bd^2}$ equal to 4.7 (actual value is 4.68).

The area of tension reinforcement required is in two parts. The first part is required to balance the compression force in the concrete, and the second part to balance the force in the compression reinforcement.

For the concrete the depth of the neutral axis is controlled by the amount of redistribution carried out. With 10%, 20% and 30% redistribution the values of x (the neutral axis depth) are $0.5d$, $0.4d$ and $0.3d$ respectively.

If we call A_b the area of reinforcement required to balance the concrete in compression, by equating forces

$$0.87 f_y\, A_b = \frac{0.67\, f_{cu} b \times 0.9\, x}{1.5}$$

From which $p_b = \dfrac{100\, A_b}{bd} = \dfrac{40.2\, f_{cu}}{400}\left(\dfrac{x}{d}\right)$ for $f_y = 460$ N/mm²

The area of compression reinforcement can be obtained by transposing the Code equation and becomes

$$p' = \frac{100\, A'_s}{bd} = \frac{100\, (R - R')}{\left(1 - \dfrac{d'}{d}\right) f_{sc}}$$

where f_{sc}, the stress in the compression reinforcement, depends on the value of d'/x. If d'/x exceeds 0.43, f_{sc} will be less than $0.87\, f_y$. With 30% redistribution carried out the stress in the compression reinforcement will be 350 N/mm² and 233 N/mm² for d'/d of 0.15 and 0.2 respectively. With 20% redistribution the stress will be 350 N/mm² for d'/d of 0.2. In all other cases in the tables the stress will be 400 N/mm².

The total area of tension reinforcement will be

$$p = \frac{100\, A_s}{bd} = p_b + \frac{p' f_{sc}}{400}$$

Minimum percentage of reinforcement

The minimum percentage of tension reinforcement in a singly reinforced section is a function of the overall depth of the section, rather than the effective depth as is used in the normal calculation.

With $100\, A_s/bh = 0.13$ the value of $100\, A_s/bd$ will vary between 0.16 for $d/h = 0.8$ to 0.14 for $d/h = 0.95$. In the tables a value of 0.15 has been used which needs to be checked for values of d/h less than 0.85.

For compression reinforcement the minimum value of $100\, A'_s/bh$ is 0.2.

A value of 0.24 has been used in the tables which again needs checking for values of d/h less than 0.85.

Values of M/bd^2 less than 0.50 require less than the minimum percentage so these values have been omitted.

Examples on use of tables

Example 1

For a rectangular section the design ultimate moment gives a value of $\dfrac{M}{bd^2}$ as 5.0. Redistribution has not been carried out and the concrete is Grade C35.

For Grade C35 the maximum value for singly reinforced is 5.46 which can be seen from Table 2. The area of reinforcement required p $\left(= \dfrac{100\,A_s}{bd} \right) = 1.56$.

Example 2

Using the data in the above example we will now assume that 30% redistribution has been carried out.

The value of R' is now 3.64 and compression reinforcement is required. Assuming $d'/d = 0.1$, from Table 7 we can find that the percentage of compression reinforcement is 0.38 and the percentage of tension reinforcement is 1.43.

Note – for values of M/bd^2 between the tabulated values, linear interpolation can be used.

SECTION A
Tables 1 – 3

Percentages of reinforcement required for beams and slabs reinforced in tension only, $f_y=460$ N/mm^2

TABLE 1
Concrete grade C30

Neutral Axis Factor x/d	Lever Arm Factor z/d	$\dfrac{M}{bd^2}$	p	Neutral Axis Factor x/d	Lever Arm Factor z/d	$\dfrac{M}{bd^2}$	p
0.111	0.950	0.50	0.15	0.114	0.949	1.30	0.34
		0.52	0.15	0.115	0.948	1.32	0.35
		0.54	0.15	0.117	0.947	1.34	0.35
		0.56	0.15	0.119	0.946	1.36	0.36
		0.58	0.15	0.121	0.946	1.38	0.37
		0.60	0.16	0.123	0.945	1.40	0.37
		0.62	0.16	0.125	0.944	1.42	0.38
		0.64	0.17	0.127	0.943	1.44	0.38
		0.66	0.17	0.128	0.942	1.46	0.39
		0.68	0.18	0.130	0.941	1.48	0.39
		0.70	0.18	0.132	0.940	1.50	0.40
		0.72	0.19	0.134	0.940	1.52	0.40
		0.74	0.20	0.136	0.939	1.54	0.41
		0.76	0.20	0.138	0.938	1.56	0.42
		0.78	0.21	0.140	0.937	1.58	0.42
		0.80	0.21	0.142	0.936	1.60	0.43
		0.82	0.22	0.144	0.935	1.62	0.43
		0.84	0.22	0.146	0.935	1.64	0.44
		0.86	0.23	0.147	0.934	1.66	0.44
		0.88	0.23	0.149	0.933	1.68	0.45
		0.90	0.24	0.151	0.932	1.70	0.46
		0.92	0.24	0.153	0.931	1.72	0.46
		0.94	0.25	0.155	0.930	1.74	0.47
		0.96	0.25	0.157	0.929	1.76	0.47
		0.98	0.26	0.159	0.928	1.78	0.48
		1.00	0.26	0.161	0.928	1.80	0.49
		1.02	0.27	0.163	0.927	1.82	0.49
		1.04	0.27	0.165	0.926	1.84	0.50
		1.06	0.28	0.167	0.925	1.86	0.50
		1.08	0.28	0.169	0.924	1.88	0.51
		1.10	0.29	0.171	0.923	1.90	0.52
		1.12	0.30	0.173	0.922	1.92	0.52
		1.14	0.30	0.175	0.921	1.94	0.53
		1.16	0.31	0.177	0.921	1.96	0.53
		1.18	0.31	0.179	0.920	1.98	0.54
		1.20	0.32	0.180	0.919	2.00	0.54
		1.22	0.32	0.182	0.918	2.02	0.55
		1.24	0.32	0.184	0.917	2.04	0.56
		1.26	0.33	0.188	0.915	2.08	0.56
0.111	0.950	1.28	0.34	0.188	0.915	2.08	0.57

TABLE 1 (Continued)
Concrete grade C30

Neutral Axis Factor x/d	Lever Arm Factor z/d	$\dfrac{M}{bd^2}$	p		Neutral Axis Factor x/d	Lever Arm Factor z/d	$\dfrac{M}{bd^2}$	p
0.190	0.914	2.10	0.57		0.274	0.877	2.90	0.83
0.192	0.913	2.12	0.58		0.277	0.876	2.92	0.83
0.194	0.912	2.14	0.59		0.279	0.875	2.94	0.84
0.196	0.912	2.16	0.59		0.281	0.874	2.96	0.85
0.198	0.911	2.18	0.60		0.283	0.873	2.98	0.85
0.201	0.910	2.20	0.61		0.285	0.872	3.00	0.86
0.203	0.909	2.22	0.61		0.288	0.871	3.02	0.87
0.205	0.908	2.24	0.62		0.290	0.870	3.04	0.87
0.207	0.907	2.26	0.62		0.292	0.869	3.06	0.88
0.209	0.906	2.28	0.63		0.294	0.868	3.08	0.89
0.211	0.905	2.30	0.64		0.297	0.867	3.10	0.89
0.213	0.904	2.32	0.64		0.299	0.865	3.12	0.90
0.215	0.903	2.34	0.65		0.301	0.864	3.14	0.91
0.217	0.902	2.36	0.65		0.303	0.863	3.16	0.92
0.219	0.901	2.38	0.66		0.306	0.862	3.18	0.92
0.221	0.901	2.40	0.67		0.308	0.861	3.20	0.93
0.223	0.900	2.42	0.67		0.310	0.860	3.22	0.94
0.225	0.899	2.44	0.68		0.313	0.859	3.24	0.94
0.227	0.898	2.46	0.69		0.315	0.858	3.26	0.95
0.229	0.897	2.48	0.69		0.317	0.857	3.28	0.96
0.231	0.896	2.50	0.70		0.320	0.856	3.30	0.96
0.233	0.895	2.52	0.70		0.322	0.855	3.32	0.97
0.236	0.894	2.54	0.71		0.324	0.854	3.34	0.98
0.238	0.893	2.56	0.72		0.327	0.853	3.36	0.99
0.240	0.892	2.58	0.72		0.329	0.852	3.38	0.99
0.242	0.891	2.60	0.73		0.331	0.851	3.40	1.00
0.244	0.890	2.62	0.74		0.334	0.850	3.42	1.01
0.246	0.889	2.64	0.74		0.336	0.849	3.44	1.01
0.248	0.888	2.66	0.75		0.338	0.848	3.46	1.02
0.250	0.887	2.68	0.76		0.341	0.847	3.48	1.03
0.253	0.886	2.70	0.76		0.343	0.846	3.50	1.04
0.255	0.885	2.72	0.77		0.346	0.844	3.52	1.04
0.257	0.884	2.74	0.78		0.348	0.843	3.54	1.05
0.259	0.883	2.76	0.78		0.350	0.842	3.56	1.06
0.261	0.882	2.78	0.79		0.353	0.841	3.58	1.06
0.263	0.881	2.80	0.79		0.355	0.840	3.60	1.07
0.266	0.880	2.82	0.80		0.358	0.839	3.62	1.08
0.268	0.880	2.84	0.81		0.360	0.838	3.64	1.09
0.270	0.878	2.86	0.81		0.363	0.837	3.66	1.09
0.272	0.878	2.88	0.82		0.365	0.836	3.68	1.10

30%

TABLE 1 (Continued)
Concrete grade C30

Neutral Axis Factor x/d	Lever Arm Factor z/d	$\dfrac{M}{bd^2}$	p		Neutral Axis Factor x/d	Lever Arm Factor z/d	$\dfrac{M}{bd^2}$	p
0.368	0.835	3.70	1.11		0.474	0.787	4.50	1.43
0.370	0.833	3.72	1.12		0.477	0.785	4.52	1.44
0.373	0.832	3.74	1.12		0.480	0.784	4.54	1.45
0.375	0.831	3.76	1.13		0.483	0.783	4.56	1.46
0.378	0.830	3.78	1.14		0.486	0.781	4.58	1.47
0.380	0.829	3.80	1.15		0.489	0.780	4.60	1.47
0.383	0.828	3.82	1.15		0.492	0.778	4.62	1.48
0.385	0.827	3.84	1.16		0.495	0.777	4.64	1.49
0.388	0.826	3.86	1.17		0.498	0.776	4.66	1.50
0.390	0.824	3.88	1.18		0.500	0.775	4.68	1.51
0.393	0.823	3.90	1.18					
0.395	0.822	3.92	1.19					
0.398	0.821	3.94	1.20					
0.401	0.820	3.96	1.21	20%				
0.403	0.819	3.98	1.22					
0.406	0.817	4.00	1.22					
0.408	0.816	4.02	1.23					
0.411	0.815	4.04	1.24					
0.414	0.814	4.06	1.25					
0.416	0.813	4.08	1.26					
0.419	0.811	4.10	1.26					
0.422	0.810	4.12	1.27					
0.424	0.809	4.14	1.28					
0.427	0.808	4.16	1.29					
0.430	0.807	4.18	1.30					
0.432	0.805	4.20	1.30					
0.435	0.804	4.22	1.31					
0.438	0.803	4.24	1.32					
0.441	0.802	4.26	1.33					
0.443	0.800	4.28	1.34					
0.446	0.799	4.30	1.35					
0.449	0.798	4.32	1.35					
0.452	0.797	4.34	1.36					
0.454	0.795	4.36	1.37					
0.457	0.794	4.38	1.38					
0.460	0.793	4.40	1.39					
0.463	0.792	4.42	1.40					
0.466	0.790	4.44	1.40					
0.469	0.789	4.46	1.41					
0.472	0.788	4.48	1.42					

TABLE 2
Concrete grade C35

Neutral Axis Factor x/d	Lever Arm Factor z/d	$\dfrac{M}{bd^2}$	p	Neutral Axis Factor x/d	Lever Arm Factor z/d	$\dfrac{M}{bd^2}$	p
0.111	0.950	0.50	0.15	0.111	0.950	1.30	0.34
		0.52	0.15			1.32	0.35
		0.54	0.15			1.34	0.35
		0.56	0.15			1.36	0.36
		0.58	0.15			1.38	0.36
		0.60	0.16			1.40	0.37
		0.62	0.16			1.42	0.37
		0.64	0.17			1.44	0.38
		0.66	0.17			1.46	0.38
		0.68	0.18	0.111	0.950	1.48	0.39
		0.70	0.18	0.112	0.949	1.50	0.39
		0.72	0.19	0.114	0.949	1.52	0.40
		0.74	0.19	0.115	0.948	1.54	0.41
		0.76	0.20	0.117	0.947	1.56	0.41
		0.78	0.21	0.119	0.947	1.58	0.42
		0.80	0.21	0.120	0.946	1.60	0.42
		0.82	0.22	0.122	0.945	1.62	0.43
		0.84	0.22	0.123	0.944	1.64	0.43
		0.86	0.23	0.125	0.944	1.66	0.44
		0.88	0.23	0.127	0.943	1.68	0.45
		0.90	0.24	0.128	0.942	1.70	0.45
		0.92	0.24	0.130	0.942	1.72	0.46
		0.94	0.25	0.131	0.941	1.74	0.46
		0.96	0.25	0.133	0.940	1.76	0.47
		0.98	0.26	0.135	0.939	1.78	0.47
		1.00	0.26	0.136	0.939	1.80	0.48
		1.02	0.27	0.138	0.938	1.82	0.49
		1.04	0.27	0.140	0.937	1.84	0.49
		1.06	0.28	0.141	0.936	1.86	0.50
		1.08	0.28	0.143	0.936	1.88	0.50
		1.10	0.29	0.144	0.935	1.90	0.51
		1.12	0.29	0.146	0.934	1.92	0.51
		1.14	0.30	0.148	0.934	1.94	0.52
		1.16	0.31	0.149	0.933	1.96	0.53
		1.18	0.31	0.151	0.932	1.98	0.53
		1.20	0.32	0.153	0.931	2.00	0.54
		1.22	0.32	0.154	0.931	2.02	0.54
		1.24	0.33	0.156	0.930	2.04	0.55
		1.26	0.33	0.158	0.929	2.06	0.55
0.111	0.950	1.28	0.34	0.159	0.928	2.08	0.56

TABLE 2 (Continued)
Concrete grade C35

Neutral Axis Factor x/d	Lever Arm Factor z/d	$\dfrac{M}{bd^2}$	p	Neutral Axis Factor x/d	Lever Arm Factor z/d	$\dfrac{M}{bd^2}$	p
0.161	0.928	2.10	0.57	0.230	0.897	2.90	0.81
0.163	0.927	2.12	0.57	0.232	0.896	2.92	0.81
0.164	0.926	2.14	0.58	0.233	0.895	2.94	0.82
0.166	0.925	2.16	0.58	0.235	0.894	2.96	0.83
0.168	0.925	2.18	0.59	0.237	0.893	2.98	0.83
0.169	0.924	2.20	0.60	0.239	0.892	3.00	0.84
0.171	0.923	2.22	0.60	0.241	0.892	3.02	0.85
0.173	0.922	2.24	0.61	0.243	0.891	3.04	0.85
0.174	0.922	2.26	0.61	0.244	0.890	3.06	0.86
0.176	0.921	2.28	0.62	0.246	0.889	3.08	0.87
0.178	0.920	2.30	0.62	0.248	0.888	3.10	0.87
0.179	0.919	2.32	0.63	0.250	0.888	3.12	0.88
0.181	0.919	2.34	0.64	0.252	0.887	3.14	0.89
0.183	0.918	2.36	0.64	0.254	0.886	3.16	0.89
0.184	0.917	2.38	0.65	0.255	0.885	3.18	0.90
0.186	0.916	2.40	0.65	0.257	0.884	3.20	0.90
0.188	0.915	2.42	0.66	0.259	0.883	3.22	0.91
0.190	0.915	2.44	0.67	0.261	0.883	3.24	0.92
0.191	0.914	2.46	0.67	0.263	0.882	3.26	0.92
0.193	0.913	2.48	0.68	0.265	0.881	3.28	0.93
0.195	0.912	2.50	0.69	0.267	0.880	3.30	0.94
0.196	0.912	2.52	0.69	0.268	0.879	3.32	0.94
0.198	0.911	2.54	0.70	0.270	0.878	3.34	0.95
0.200	0.910	2.56	0.70	0.272	0.878	3.36	0.96
0.202	0.909	2.58	0.71	0.274	0.877	3.38	0.96
0.203	0.908	2.60	0.72	0.276	0.876	3.40	0.97
0.205	0.908	2.62	0.72	0.278	0.875	3.42	0.98
0.207	0.907	2.64	0.73	0.280	0.874	3.44	0.98
0.209	0.906	2.66	0.73	0.282	0.873	3.46	0.99
0.210	0.905	2.68	0.74	0.284	0.872	3.48	1.00
0.212	0.905	2.70	0.75	0.285	0.872	3.50	1.00
0.214	0.904	2.72	0.75	0.287	0.871	3.52	1.01
0.216	0.903	2.74	0.76	0.289	0.870	3.54	1.02
0.217	0.902	2.76	0.76	0.291	0.869	3.56	1.02
0.219	0.901	2.78	0.77	0.293	0.868	3.58	1.03
0.221	0.901	2.80	0.78	0.295	0.867	3.60	1.04
0.223	0.900	2.82	0.78	0.297	0.866	3.62	1.04
0.225	0.899	2.84	0.79	0.299	0.865	3.64	1.05	30%
0.226	0.898	2.86	0.80	0.301	0.865	3.66	1.06
0.228	0.897	2.88	0.80	0.303	0.864	3.68	1.07

TABLE 2 (Continued)
Concrete grade C35

Neutral Axis Factor x/d	Lever Arm Factor z/d	$\dfrac{M}{bd^2}$	p		Neutral Axis Factor x/d	Lever Arm Factor z/d	$\dfrac{M}{bd^2}$	p	
0.305	0.863	3.70	1.07		0.387	0.826	4.50	1.36	
0.307	0.862	3.72	1.08		0.390	0.825	4.52	1.37	
0.309	0.861	3.74	1.09		0.392	0.824	4.54	1.38	
0.311	0.860	3.76	1.09		0.394	0.823	4.56	1.39	
0.313	0.859	3.78	1.10		0.396	0.822	4.58	1.39	
0.315	0.858	3.80	1.11		0.398	0.821	4.60	1.40	
0.317	0.857	3.82	1.11		0.401	0.820	4.62	1.41	20%
0.319	0.857	3.84	1.12		0.403	0.819	4.64	1.42	
0.321	0.856	3.86	1.13		0.405	0.818	4.66	1.42	
0.323	0.855	3.88	1.13		0.407	0.817	4.68	1.43	
0.325	0.854	3.90	1.14		0.410	0.816	4.70	1.44	
0.327	0.853	3.92	1.15		0.412	0.815	4.72	1.45	
0.329	0.852	3.94	1.16		0.414	0.814	4.74	1.46	
0.331	0.851	3.96	1.16		0.416	0.813	4.76	1.46	
0.333	0.850	3.98	1.17		0.419	0.812	4.78	1.47	
0.335	0.849	4.00	1.18		0.421	0.811	4.80	1.48	
0.337	0.848	4.02	1.18		0.423	0.810	4.82	1.49	
0.339	0.848	4.04	1.19		0.425	0.809	4.84	1.50	
0.341	0.847	4.06	1.20		0.428	0.808	4.86	1.50	
0.343	0.846	4.08	1.21		0.430	0.806	4.88	1.51	
0.345	0.845	4.10	1.21		0.432	0.805	4.90	1.52	
0.347	0.844	4.12	1.22		0.435	0.804	4.92	1.53	
0.349	0.843	4.14	1.23		0.437	0.803	4.94	1.54	
0.351	0.842	4.16	1.24		0.439	0.802	4.96	1.55	
0.353	0.841	4.18	1.24		0.442	0.801	4.98	1.55	
0.355	0.840	4.20	1.25		0.444	0.800	5.00	1.56	
0.357	0.839	4.22	1.26		0.447	0.799	5.02	1.57	
0.360	0.838	4.24	1.26		0.449	0.798	5.04	1.58	
0.362	0.837	4.26	1.27		0.451	0.797	5.06	1.59	
0.364	0.836	4.28	1.28		0.454	0.796	5.08	1.60	
0.366	0.835	4.30	1.29		0.456	0.795	5.10	1.60	
0.368	0.834	4.32	1.29		0.458	0.794	5.12	1.61	
0.370	0.833	4.34	1.30		0.461	0.793	5.14	1.62	
0.372	0.832	4.36	1.31		0.463	0.791	5.16	1.63	
0.374	0.832	4.38	1.32		0.466	0.790	5.18	1.64	
0.377	0.831	4.40	1.32		0.468	0.789	5.20	1.65	
0.379	0.830	4.42	1.33		0.471	0.788	5.22	1.66	
0.381	0.829	4.44	1.34		0.473	0.787	5.24	1.66	
0.383	0.828	4.46	1.35		0.476	0.786	5.26	1.67	
0.385	0.827	4.48	1.35		0.478	0.785	5.28	1.68	

TABLE 2 (Continued)
Concrete grade C35

Neutral Axis Factor x/d	Lever Arm Factor z/d	$\dfrac{M}{bd^2}$	p
0.481	0.784	5.30	1.69
0.483	0.783	5.32	1.70
0.486	0.781	5.34	1.71
0.488	0.780	5.36	1.72
0.491	0.779	5.38	1.73
0.493	0.778	5.40	1.74
0.496	0.777	5.42	1.74
0.498	0.776	5.44	1.75
0.500	0.775	5.46	1.76

TABLE 3
Concrete grade C40

Neutral Axis Factor x/d	Lever Arm Factor z/d	$\dfrac{M}{bd^2}$	p	Neutral Axis Factor x/d	Lever Arm Factor z/d	$\dfrac{M}{bd^2}$	p
0.111	0.950	0.50	0.15	0.111	0.950	1.30	0.34
		0.52	0.15			1.32	0.35
		0.54	0.15			1.34	0.35
		0.56	0.15			1.36	0.36
		0.58	0.15			1.38	0.36
		0.60	0.16			1.40	0.37
		0.62	0.16			1.42	0.37
		0.64	0.17			1.44	0.38
		0.66	0.17			1.46	0.38
		0.68	0.18			1.48	0.39
		0.70	0.18			1.50	0.39
		0.72	0.19			1.52	0.40
		0.74	0.19			1.54	0.41
		0.76	0.20			1.56	0.41
		0.78	0.21			1.58	0.42
		0.80	0.21			1.60	0.42
		0.82	0.22			1.62	0.43
		0.84	0.22			1.64	0.43
		0.86	0.23			1.66	0.44
		0.88	0.23			1.68	0.44
		0.90	0.24	0.111	0.950	1.70	0.45
		0.92	0.24	0.113	0.949	1.72	0.45
		0.94	0.25	0.114	0.949	1.74	0.46
		0.96	0.25	0.115	0.948	1.76	0.46
		0.98	0.26	0.117	0.947	1.78	0.47
		1.00	0.26	0.118	0.947	1.80	0.48
		1.02	0.27	0.120	0.946	1.82	0.48
		1.04	0.27	0.121	0.946	1.84	0.49
		1.06	0.28	0.122	0.945	1.86	0.49
		1.08	0.28	0.124	0.944	1.88	0.50
		1.10	0.29	0.125	0.944	1.90	0.50
		1.12	0.29	0.127	0.943	1.92	0.51
		1.14	0.30	0.128	0.942	1.94	0.51
		1.16	0.31	0.129	0.942	1.96	0.52
		1.18	0.31	0.131	0.941	1.98	0.53
		1.20	0.32	0.132	0.940	2.00	0.53
		1.22	0.32	0.134	0.940	2.02	0.54
		1.24	0.33	0.135	0.939	2.04	0.54
		1.26	0.33	0.136	0.939	2.06	0.55
0.111	0.950	1.28	0.34	0.138	0.938	2.08	0.55

TABLE 3 (Continued)
Concrete grade C40

Neutral Axis Factor x/d	Lever Arm Factor z/d	$\dfrac{M}{bd^2}$	p	Neutral Axis Factor x/d	Lever Arm Factor z/d	$\dfrac{M}{bd^2}$	p
0.139	0.937	2.10	0.56	0.198	0.911	2.90	0.80
0.141	0.937	2.12	0.57	0.200	0.910	2.92	0.80
0.142	0.936	2.14	0.57	0.201	0.910	2.94	0.81
0.144	0.935	2.16	0.58	0.203	0.909	2.96	0.81
0.145	0.935	2.18	0.58	0.204	0.908	2.98	0.82
0.146	0.934	2.20	0.59	0.206	0.907	3.00	0.83
0.148	0.933	2.22	0.59	0.207	0.907	3.02	0.83
0.149	0.933	2.24	0.60	0.209	0.906	3.04	0.84
0.151	0.932	2.26	0.61	0.210	0.905	3.06	0.84
0.152	0.932	2.28	0.61	0.212	0.905	3.08	0.85
0.154	0.931	2.30	0.62	0.213	0.904	3.10	0.86
0.155	0.930	2.32	0.62	0.215	0.903	3.12	0.86
0.157	0.930	2.34	0.63	0.216	0.903	3.14	0.87
0.158	0.929	2.36	0.64	0.218	0.902	3.16	0.88
0.159	0.928	2.38	0.64	0.219	0.901	3.18	0.88
0.161	0.928	2.40	0.65	0.221	0.901	3.20	0.89
0.162	0.927	2.42	0.65	0.223	0.900	3.22	0.89
0.164	0.926	2.44	0.66	0.224	0.899	3.24	0.90
0.165	0.926	2.46	0.66	0.226	0.898	3.26	0.91
0.167	0.925	2.48	0.67	0.227	0.898	3.28	0.91
0.168	0.924	2.50	0.68	0.229	0.897	3.30	0.92
0.170	0.924	2.52	0.68	0.230	0.896	3.32	0.93
0.171	0.923	2.54	0.69	0.232	0.896	3.34	0.93
0.173	0.922	2.56	0.69	0.233	0.895	3.36	0.94
0.174	0.922	2.58	0.70	0.235	0.894	3.38	0.94
0.176	0.921	2.60	0.71	0.237	0.894	3.40	0.95
0.177	0.920	2.62	0.71	0.238	0.893	3.42	0.96
0.179	0.920	2.64	0.72	0.240	0.892	3.44	0.96
0.180	0.919	2.66	0.72	0.241	0.891	3.46	0.97
0.181	0.918	2.68	0.73	0.243	0.891	3.48	0.98
0.183	0.918	2.70	0.74	0.245	0.890	3.50	0.98
0.184	0.917	2.72	0.74	0.246	0.889	3.52	0.99
0.186	0.916	2.74	0.75	0.248	0.889	3.54	1.00
0.187	0.916	2.76	0.75	0.249	0.888	3.56	1.00
0.189	0.915	2.78	0.76	0.251	0.887	3.58	1.01
0.190	0.914	2.80	0.77	0.253	0.886	3.60	1.02
0.192	0.914	2.82	0.77	0.254	0.886	3.62	1.02
0.193	0.913	2.84	0.78	0.256	0.885	3.64	1.03
0.195	0.912	2.86	0.78	0.257	0.884	3.66	1.03
0.196	0.912	2.88	0.79	0.259	0.883	3.68	1.04

TABLE 3 (Continued)
Concrete grade C40

Neutral Axis Factor x/d	Lever Arm Factor z/d	$\dfrac{M}{bd^2}$	p	Neutral Axis Factor x/d	Lever Arm Factor z/d	$\dfrac{M}{bd^2}$	p	
0.261	0.883	3.70	1.05	0.328	0.852	4.50	1.32	
0.262	0.882	3.72	1.05	0.330	0.851	4.52	1.33	
0.264	0.881	3.74	1.06	0.332	0.851	4.54	1.33	
0.266	0.880	3.76	1.07	0.334	0.850	4.56	1.34	
0.267	0.880	3.78	1.07	0.335	0.849	4.58	1.35	
0.269	0.879	3.80	1.08	0.337	0.848	4.60	1.36	
0.270	0.878	3.82	1.09	0.339	0.847	4.62	1.36	
0.272	0.878	3.84	1.09	0.341	0.847	4.64	1.37	
0.274	0.877	3.86	1.10	0.343	0.846	4.66	1.38	
0.275	0.876	3.88	1.11	0.344	0.845	4.68	1.38	
0.277	0.875	3.90	1.11	0.346	0.844	4.70	1.39	
0.279	0.875	3.92	1.12	0.348	0.843	4.72	1.40	
0.280	0.874	3.94	1.13	0.350	0.843	4.74	1.41	
0.282	0.873	3.96	1.13	0.352	0.842	4.76	1.41	
0.284	0.872	3.98	1.14	0.353	0.841	4.78	1.42	
0.285	0.872	4.00	1.15	0.355	0.840	4.80	1.43	
0.287	0.871	4.02	1.15	0.357	0.839	4.82	1.44	
0.289	0.870	4.04	1.16	0.359	0.838	4.84	1.44	
0.290	0.869	4.06	1.17	0.361	0.838	4.86	1.45	
0.292	0.869	4.08	1.17	0.363	0.837	4.88	1.46	
0.294	0.868	4.10	1.18	0.365	0.836	4.90	1.47	
0.296	0.867	4.12	1.19	0.366	0.835	4.92	1.47	
0.297	0.866	4.14	1.19	0.368	0.834	4.94	1.48	
0.299	0.865	4.16	1.20	0.370	0.833	4.96	1.49	30%
0.301	0.865	4.18	1.21	0.372	0.833	4.98	1.50	
0.302	0.864	4.20	1.22	0.374	0.832	5.00	1.50	
0.304	0.863	4.22	1.22	0.376	0.831	5.02	1.51	
0.306	0.862	4.24	1.23	0.378	0.830	5.04	1.52	
0.307	0.862	4.26	1.24	0.379	0.829	5.06	1.53	
0.309	0.861	4.28	1.24	0.381	0.828	5.08	1.53	
0.311	0.860	4.30	1.25	0.383	0.828	5.10	1.54	
0.313	0.859	4.32	1.26	0.385	0.827	5.12	1.55	
0.314	0.859	4.34	1.26	0.387	0.826	5.14	1.56	
0.316	0.858	4.36	1.27	0.389	0.825	5.16	1.56	
0.318	0.857	4.38	1.28	0.391	0.824	5.18	1.57	
0.320	0.856	4.40	1.28	0.393	0.823	5.20	1.58	
0.321	0.855	4.42	1.29	0.395	0.822	5.22	1.59	
0.323	0.855	4.44	1.30	0.397	0.821	5.24	1.59	
0.325	0.854	4.46	1.31	0.399	0.821	5.26	1.60	
0.327	0.853	4.48	1.31	0.401	0.820	5.28	1.61	20%

TABLE 3 (Continued)
Concrete grade C40

Neutral Axis Factor x/d	Lever Arm Factor z/d	$\frac{M}{bd^2}$	p	Neutral Axis Factor x/d	Lever Arm Factor z/d	$\frac{M}{bd^2}$	p
0.403	0.819	5.30	1.62	0.485	0.782	6.10	1.95
0.404	0.818	5.32	1.63	0.488	0.781	6.12	1.96
0.406	0.817	5.34	1.63	0.490	0.780	6.14	1.97
0.408	0.816	5.36	1.64	0.492	0.779	6.16	1.98
0.410	0.815	5.38	1.65	0.494	0.778	6.18	1.99
0.412	0.814	5.40	1.66	0.497	0.777	6.20	2.00
0.414	0.814	5.42	1.67	0.499	0.776	6.22	2.01
0.416	0.813	5.44	1.67	0.500	0.775	6.24	2.01
0.418	0.812	5.46	1.68				
0.420	0.811	5.48	1.69				
0.422	0.810	5.50	1.70				
0.424	0.809	5.52	1.71				
0.426	0.808	5.54	1.71				
0.428	0.807	5.56	1.72				
0.430	0.806	5.58	1.73				
0.432	0.805	5.60	1.74				
0.434	0.805	5.62	1.75				
0.436	0.804	5.64	1.75				
0.439	0.803	5.66	1.76				
0.441	0.802	5.68	1.77				
0.443	0.801	5.70	1.78				
0.445	0.800	5.72	1.79				
0.447	0.799	5.74	1.80				
0.449	0.798	5.76	1.80				
0.451	0.797	5.78	1.81				
0.453	0.796	5.80	1.82				
0.455	0.795	5.82	1.83				
0.457	0.794	5.84	1.84				
0.459	0.793	5.86	1.85				
0.462	0.792	5.90	1.86				
0.464	0.791	5.90	1.86				
0.466	0.790	5.92	1.87				
0.468	0.789	5.94	1.88				
0.470	0.788	5.96	1.89				
0.472	0.787	5.98	1.90				
0.474	0.787	6.00	1.91				
0.477	0.786	6.02	1.92				
0.479	0.786	6.04	1.92				
0.481	0.784	6.06	1.93				
0.483	0.783	6.08	1.94				

Percentages of reinforcement required for beams and slabs reinforced in both tension and compression, $f_y = 460 \, \text{N/mm}^2$

TABLE 4
Concrete grade C30

$$d'/d = 0.10$$

Redist.	≤ 10%		20%		30%	
$\frac{M}{bd^2}$	p'	p	p'	p	p'	p
3.2					0.24	0.93
3.3					0.24	0.95
3.4			Singly		0.24	0.98
3.5					0.24	1.01
3.6			Reinforced		0.24	1.04
3.7	Singly				0.24	1.07
3.8					0.24	1.09
3.9	Reinforced				0.24	1.12
4.0			0.24	1.22	0.24	1.15
4.1			0.24	1.24	0.27	1.18
4.2			0.24	1.27	0.30	1.20
4.3			0.24	1.30	0.33	1.23
4.4			0.24	1.33	0.36	1.26
4.5			0.24	1.36	0.38	1.29
4.6			0.24	1.38	0.41	1.32
4.7	0.24	1.51	0.24	1.41	0.44	1.34
4.8	0.24	1.54	0.24	1.44	0.47	1.37
4.9	0.24	1.57	0.26	1.47	0.49	1.40
5.0	0.24	1.60	0.29	1.49	0.52	1.43
5.1	0.24	1.62	0.32	1.52	0.55	1.45
5.2	0.24	1.65	0.34	1.55	0.58	1.48
5.3	0.24	1.68	0.37	1.58	0.61	1.51
5.4	0.24	1.71	0.40	1.61	0.63	1.54
5.5	0.24	1.74	0.43	1.63	0.66	1.57
5.6	0.26	1.76	0.46	1.66	0.69	1.59
5.7	0.28	1.79	0.48	1.69	0.72	1.62
5.8	0.31	1.82	0.51	1.72	0.74	1.65
5.9	0.34	1.85	0.54	1.74	0.77	1.68
6.0	0.37	1.87	0.57	1.77	0.80	1.70
6.1	0.39	1.90	0.59	1.80	0.83	1.73
6.2	0.42	1.93	0.62	1.83	0.86	1.76
6.3	0.45	1.96	0.65	1.86	0.88	1.79
6.4	0.48	1.99	0.68	1.88	0.91	1.82
6.5	0.51	2.01	0.71	1.91	0.94	1.84
6.6	0.53	2.04	0.73	1.94	0.97	1.87
6.7	0.56	2.07	0.76	1.97	0.99	1.90
6.8	0.59	2.10	0.79	1.99	1.02	1.93
6.9	0.62	2.12	0.82	2.02	1.05	1.95
7.0	0.64	2.15	0.84	2.05	1.08	1.98
7.1	0.67	2.18	0.87	2.08	1.11	2.01
7.2	0.70	2.21	0.90	2.11	1.13	2.04
7.3	0.73	2.24	0.93	2.13	1.16	2.07
7.4	0.76	2.26	0.96	2.16	1.19	2.09

TABLE 4 (Continued)
Concrete grade C30

$d'/d = 0.10$

Redist.	≤ 10%		20%		30%	
$\frac{M}{bd^2}$	p'	p	p'	p	p'	p
7.5	0.78	2.29	0.98	2.19	1.22	2.12
7.6	0.81	2.32	1.01	2.22	1.24	2.15
7.7	0.84	2.35	1.04	2.24	1.27	2.18
7.8	0.87	2.37	1.07	2.27	1.30	2.20
7.9	0.89	2.40	1.09	2.30	1.33	2.23
8.0	0.92	2.43	1.12	2.33	1.36	2.26
8.1	0.95	2.46	1.15	2.36	1.38	2.29
8.2	0.98	2.49	1.18	2.38	1.41	2.32
8.3	1.01	2.51	1.21	2.41	1.44	2.34
8.4	1.03	2.54	1.23	2.44	1.47	2.37
8.5	1.06	2.57	1.26	2.47	1.49	2.40
8.6	1.09	2.60	1.29	2.49	1.52	2.43
8.7	1.12	2.62	1.32	2.52	1.55	2.45
8.8	1.14	2.65	1.34	2.55	1.58	2.48
8.9	1.17	2.68	1.37	2.58	1.61	2.51
9.0	1.20	2.71	1.40	2.61	1.63	2.54
9.1	1.23	2.74	1.43	2.63	1.66	2.57
9.2	1.26	2.76	1.46	2.66	1.69	2.59
9.3	1.28	2.79	1.48	2.69	1.72	2.62
9.4	1.31	2.82	1.51	2.72	1.74	2.65
9.5	1.34	2.85	1.54	2.74	1.77	2.68
9.6	1.37	2.87	1.57	2.77	1.80	2.70
9.7	1.39	2.90	1.59	2.80	1.83	2.73
9.8	1.42	2.93	1.62	2.83	1.86	2.76
9.9	1.45	2.96	1.65	2.86	1.88	2.79
10.0	1.48	2.99	1.68	2.88	1.91	2.82
10.1	1.51	3.01	1.71	2.91	1.94	2.84
10.2	1.53	3.04	1.73	2.94	1.97	2.87
10.3	1.56	3.07	1.76	2.97	1.99	2.90
10.4	1.59	3.10	1.79	2.99	2.02	2.93
10.5	1.62	3.12	1.82	3.02	2.05	2.95
10.6	1.64	3.15	1.84	3.05	2.08	2.98
10.7	1.67	3.18	1.87	3.08	2.11	3.01
10.8	1.70	3.21	1.90	3.11	2.13	3.04
10.9	1.73	3.24	1.93	3.13	2.16	3.07
11.0	1.76	3.26	1.96	3.16	2.19	3.09
11.1	1.78	3.29	1.98	3.19	2.22	3.12
11.2	1.81	3.32	2.01	3.22	2.24	3.15
11.3	1.84	3.35	2.04	3.24	2.27	3.18
11.4	1.87	3.37	2.07	3.27	2.30	3.20
11.5	1.89	3.40	2.09	3.30	2.33	3.23
11.6	1.92	3.43	2.12	3.33	2.36	3.26
11.7	1.95	3.46	2.15	3.36	2.38	3.29

TABLE 5
Concrete grade C30

$$d'/d = 0.15$$

Redist.	≤ 10%		20%		30%	
$\dfrac{M}{bd^2}$	p'	p	p'	p	p'	p
3.2					0.24	0.93
3.3					0.24	0.96
3.4			Singly		0.24	0.99
3.5					0.24	1.02
3.6			Reinforced		0.24	1.05
3.7	Singly				0.24	1.08
3.8					0.24	1.10
3.9	Reinforced				0.26	1.13
4.0			0.24	1.22	0.30	1.16
4.1			0.24	1.25	0.33	1.19
4.2			0.24	1.28	0.36	1.22
4.3			0.24	1.31	0.40	1.25
4.4			0.24	1.34	0.43	1.28
4.5			0.24	1.36	0.46	1.31
4.6			0.24	1.39	0.50	1.34
4.7	0.24	1.51	0.24	1.42	0.53	1.37
4.8	0.24	1.54	0.25	1.45	0.56	1.40
4.9	0.24	1.57	0.28	1.48	0.60	1.43
5.0	0.24	1.60	0.31	1.51	0.63	1.46
5.1	0.24	1.63	0.34	1.54	0.67	1.49
5.2	0.24	1.66	0.36	1.57	0.70	1.52
5.3	0.24	1.69	0.39	1.60	0.73	1.55
5.4	0.24	1.72	0.42	1.63	0.77	1.58
5.5	0.24	1.75	0.45	1.66	0.80	1.60
5.6	0.27	1.78	0.48	1.69	0.83	1.63
5.7	0.30	1.81	0.51	1.72	0.87	1.66
5.8	0.33	1.84	0.54	1.75	0.90	1.69
5.9	0.36	1.87	0.57	1.78	0.93	1.72
6.0	0.39	1.90	0.60	1.81	0.97	1.75
6.1	0.42	1.93	0.63	1.84	1.00	1.78
6.2	0.45	1.95	0.66	1.86	1.04	1.81
6.3	0.48	1.98	0.69	1.89	1.07	1.84
6.4	0.51	2.01	0.72	1.92	1.10	1.87
6.5	0.54	2.04	0.75	1.95	1.14	1.90
6.6	0.56	2.07	0.78	1.98	1.17	1.93
6.7	0.59	2.10	0.81	2.01	1.20	1.96
6.8	0.62	2.13	0.84	2.04	1.24	1.99
6.9	0.65	2.16	0.86	2.07	1.27	2.02
7.0	0.68	2.19	0.89	2.10	1.30	2.05
7.1	0.71	2.22	0.92	2.13	1.34	2.08
7.2	0.74	2.25	0.95	2.16	1.37	2.10
7.3	0.77	2.28	0.98	2.19	1.41	2.13
7.4	0.80	2.31	1.01	2.22	1.44	2.16

TABLE 5 (Continued)
Concrete grade C30

$$d'/d = 0.15$$

Redist.	≤ 10%		20%		30%	
$\dfrac{M}{bd^2}$	p'	p	p'	p	p'	p
7.5	0.83	2.34	1.04	2.25	1.47	2.19
7.6	0.86	2.37	1.07	2.28	1.51	2.22
7.7	0.89	2.40	1.10	2.31	1.54	2.25
7.8	0.92	2.43	1.13	2.34	1.57	2.28
7.9	0.95	2.45	1.16	2.36	1.61	2.31
8.0	0.98	2.48	1.19	2.39	1.64	2.34
8.1	1.01	2.51	1.22	2.42	1.67	2.37
8.2	1.04	2.54	1.25	2.45	1.71	2.40
8.3	1.06	2.57	1.28	2.48	1.74	2.43
8.4	1.09	2.60	1.31	2.51	1.77	2.46
8.5	1.12	2.63	1.34	2.54	1.81	2.49
8.6	1.15	2.66	1.36	2.57	1.84	2.52
8.7	1.18	2.69	1.39	2.60	1.88	2.55
8.8	1.21	2.72	1.42	2.63	1.91	2.58
8.9	1.24	2.75	1.45	2.66	1.94	2.60
9.0	1.27	2.78	1.48	2.69	1.98	2.63
9.1	1.30	2.81	1.51	2.72	2.01	2.66
9.2	1.33	2.84	1.54	2.75	2.04	2.69
9.3	1.36	2.87	1.57	2.78	2.08	2.72
9.4	1.39	2.90	1.60	2.81	2.11	2.75
9.5	1.42	2.93	1.63	2.84	2.14	2.78
9.6	1.45	2.95	1.66	2.86	2.18	2.81
9.7	1.48	2.98	1.69	2.89	2.21	2.84
9.8	1.51	3.01	1.72	2.92	2.24	2.87
9.9	1.54	3.04	1.75	2.95	2.28	2.90
10.0	1.56	3.07	1.78	2.98	2.31	2.93
10.1	1.59	3.10	1.81	3.01	2.35	2.96
10.2	1.62	3.13	1.84	3.04	2.38	2.99
10.3	1.65	3.16	1.86	3.07	2.41	3.02
10.4	1.68	3.19	1.89	3.10	2.45	3.05
10.5	1.71	3.22	1.92	3.13	2.48	3.08
10.6	1.74	3.25	1.95	3.16	2.51	3.10
10.7	1.77	3.28	1.98	3.19	2.55	3.13
10.8	1.80	3.31	2.01	3.22	2.58	3.16
10.9	1.83	3.34	2.04	3.25	2.62	3.19
11.0	1.86	3.37	2.07	3.28	2.65	3.22
11.1	1.89	3.40	2.10	3.31	2.68	3.25
11.2	1.92	3.43	2.13	3.34	2.72	3.28
11.3	1.95	3.45	2.16	3.36	2.75	3.31
11.4	1.98	3.48	2.19	3.39	2.78	3.34
11.5	2.01	3.51	2.22	3.42	2.82	3.37
11.6	2.04	3.54	2.25	3.45	2.85	3.40
11.7	2.06	3.57	2.28	3.48	2.88	3.43

TABLE 6
Concrete grade C30

$$d'/d = 0.20$$

Redist.	≤ 10%		20%		30%	
$\dfrac{M}{bd^2}$	p'	p	p'	p	p'	p
3.2					0.24	0.93
3.3					0.24	0.96
3.4			Singly		0.24	0.99
3.5					0.24	1.02
3.6			Reinforced		0.26	1.05
3.7	Singly				0.31	1.09
3.8					0.36	1.12
3.9	Reinforced				0.42	1.15
4.0			0.24	1.22	0.47	1.18
4.1			0.24	1.25	0.53	1.21
4.2			0.24	1.28	0.58	1.24
4.3			0.24	1.31	0.63	1.27
4.4			0.24	1.34	0.69	1.30
4.5			0.24	1.37	0.74	1.34
4.6			0.24	1.41	0.79	1.37
4.7	0.24	1.51	0.26	1.44	0.85	1.40
4.8	0.24	1.55	0.30	1.47	0.90	1.43
4.9	0.24	1.58	0.34	1.50	0.95	1.46
5.0	0.24	1.61	0.37	1.53	1.01	1.49
5.1	0.24	1.64	0.41	1.56	1.06	1.52
5.2	0.24	1.67	0.44	1.59	1.12	1.55
5.3	0.24	1.70	0.48	1.62	1.17	1.59
5.4	0.24	1.73	0.51	1.66	1.22	1.62
5.5	0.26	1.76	0.55	1.69	1.28	1.65
5.6	0.29	1.80	0.59	1.72	1.33	1.68
5.7	0.32	1.83	0.62	1.75	1.38	1.71
5.8	0.35	1.86	0.66	1.78	1.44	1.74
5.9	0.38	1.89	0.69	1.81	1.49	1.77
6.0	0.41	1.92	0.73	1.84	1.55	1.80
6.1	0.44	1.95	0.76	1.87	1.60	1.84
6.2	0.48	1.98	0.80	1.91	1.65	1.87
6.3	0.51	2.01	0.84	1.94	1.71	1.90
6.4	0.54	2.05	0.87	1.97	1.76	1.93
6.5	0.57	2.08	0.91	2.00	1.81	1.96
6.6	0.60	2.11	0.94	2.03	1.87	1.99
6.7	0.63	2.14	0.98	2.06	1.92	2.02
6.8	0.66	2.17	1.01	2.09	1.97	2.05
6.9	0.69	2.20	1.05	2.12	2.03	2.09
7.0	0.73	2.23	1.09	2.16	2.08	2.12
7.1	0.76	2.26	1.12	2.19	2.14	2.15
7.2	0.79	2.30	1.16	2.22	2.19	2.18
7.3	0.82	2.33	1.19	2.25	2.24	2.21
7.4	0.85	2.36	1.23	2.28	2.30	2.24

TABLE 6 (Continued)
Concrete grade C30

$$d'/d = 0.20$$

Redist. $\dfrac{M}{bd^2}$	≤ 10%		20%		30%	
	p'	p	p'	p	p'	p
7.5	0.88	2.39	1.26	2.31	2.35	2.27
7.6	0.91	2.42	1.30	2.34	2.40	2.30
7.7	0.94	2.45	1.34	2.37	2.46	2.34
7.8	0.98	2.48	1.37	2.41	2.51	2.37
7.9	1.01	2.51	1.41	2.44	2.56	2.40
8.0	1.04	2.55	1.44	2.47	2.62	2.43
8.1	1.07	2.58	1.48	2.50	2.67	2.46
8.2	1.10	2.61	1.51	2.53	2.73	2.49
8.3	1.13	2.64	1.55	2.56	2.78	2.52
8.4	1.16	2.67	1.59	2.59	2.83	2.55
8.5	1.19	2.70	1.62	2.62	2.89	2.59
8.6	1.23	2.73	1.66	2.66	2.94	2.62
8.7	1.26	2.76	1.69	2.69	2.99	2.65
8.8	1.29	2.80	1.73	2.72	3.05	2.68
8.9	1.32	2.83	1.76	2.75	3.10	2.71
9.0	1.35	2.86	1.80	2.78	3.15	2.74
9.1	1.38	2.89	1.84	2.81	3.21	2.77
9.2	1.41	2.92	1.87	2.84	3.26	2.80
9.3	1.44	2.95	1.91	2.87	3.32	2.84
9.4	1.48	2.98	1.94	2.91	3.37	2.87
9.5	1.51	3.01	1.98	2.94	3.42	2.90
9.6	1.54	3.05	2.01	2.97	3.48	2.93
9.7	1.57	3.08	2.05	3.00	3.53	2.96
9.8	1.60	3.11	2.09	3.03	3.58	2.99
9.9	1.63	3.14	2.12	3.06	3.64	3.02
10.0	1.66	3.17	2.16	3.09	3.69	3.05
10.1	1.69	3.20	2.19	3.12	3.74	3.09
10.2	1.73	3.23	2.23	3.16	3.80	3.12
10.3	1.76	3.26	2.26	3.19	3.85	3.15
10.4	1.79	3.30	2.30	3.22	3.91	3.18
10.5	1.82	3.33	2.34	3.25	3.96	3.21
10.6	1.85	3.36	2.37	3.28	4.01	3.24
10.7	1.88	3.39	2.41	3.31	4.07	3.27
10.8	1.91	3.42	2.44	3.34	4.12	3.30
10.9	1.94	3.45	2.48	3.37	4.17	3.34
11.0	1.98	3.48	2.51	3.41	4.23	3.37
11.1	2.01	3.51	2.55	3.44	4.28	3.40
11.2	2.04	3.55	2.59	3.47	4.33	3.43
11.3	2.07	3.58	2.62	3.50	4.39	3.46
11.4	2.10	3.61	2.66	3.53	4.44	3.49
11.5	2.13	3.64	2.69	3.56	4.50	3.52
11.6	2.16	3.67	2.73	3.59	4.55	3.55
11.7	2.19	3.70	2.76	3.62	4.60	3.59

TABLE 7
Concrete grade C35

$$d'/d = 0.10$$

Redist. $\dfrac{M}{bd^2}$	≤ 10%		20%		30%	
	p'	p	p'	p	p'	p
3.7					0.24	1.07
3.8					0.24	1.10
3.9					0.24	1.13
4.0					0.24	1.16
4.1			Singly Reinforced		0.24	1.18
4.2		Singly Reinforced			0.24	1.21
4.3					0.24	1.24
4.4					0.26	1.27
4.5					0.26	1.29
4.6					0.27	1.32
4.7			0.24	1.43	0.29	1.35
4.8			0.24	1.46	0.32	1.38
4.9			0.24	1.48	0.35	1.41
5.0			0.24	1.51	0.38	1.43
5.1			0.24	1.54	0.41	1.46
5.2			0.24	1.57	0.43	1.49
5.3			0.24	1.60	0.46	1.52
5.4			0.24	1.62	0.49	1.54
5.5	0.24	1.77	0.24	1.65	0.52	1.57
5.6	0.24	1.80	0.27	1.68	0.54	1.60
5.7	0.24	1.83	0.30	1.71	0.57	1.63
5.8	0.24	1.85	0.33	1.73	0.60	1.66
5.9	0.24	1.88	0.36	1.76	0.63	1.68
6.0	0.24	1.91	0.38	1.79	0.66	1.71
6.1	0.24	1.94	0.41	1.82	0.68	1.74
6.2	0.24	1.96	0.44	1.85	0.71	1.77
6.3	0.24	1.99	0.47	1.87	0.74	1.79
6.4	0.26	2.02	0.49	1.90	0.77	1.82
6.5	0.29	2.05	0.52	1.93	0.79	1.85
6.6	0.32	2.08	0.55	1.96	0.82	1.88
6.7	0.34	2.10	0.58	1.98	0.85	1.91
6.8	0.37	2.13	0.61	2.01	0.88	1.93
6.9	0.40	2.16	0.63	2.04	0.91	1.96
7.0	0.43	2.19	0.66	2.07	0.93	1.99
7.1	0.46	2.21	0.69	2.10	0.96	2.02
7.2	0.48	2.24	0.72	2.12	0.99	2.04
7.3	0.51	2.27	0.74	2.15	1.02	2.07
7.4	0.54	2.30	0.77	2.18	1.04	2.10
7.5	0.57	2.33	0.80	2.21	1.07	2.13
7.6	0.59	2.35	0.83	2.23	1.10	2.16
7.7	0.62	2.38	0.86	2.26	1.13	2.18
7.8	0.65	2.41	0.88	2.29	1.16	2.21
7.9	0.68	2.44	0.91	2.32	1.18	2.24

TABLE 7 (Continued)
Concrete grade C35

$$d'/d = 0.10$$

Redist.	≤ 10%		20%		30%	
$\dfrac{M}{bd^2}$	p'	p	p'	p	p'	p
8.0	0.71	2.46	0.94	2.35	1.21	2.27
8.1	0.73	2.49	0.97	2.37	1.24	2.29
8.2	0.76	2.52	0.99	2.40	1.27	2.32
8.3	0.79	2.55	1.02	2.43	1.29	2.35
8.4	0.82	2.58	1.05	2.46	1.32	2.38
8.5	0.84	2.60	1.08	2.48	1.35	2.41
8.6	0.87	2.63	1.11	2.51	1.38	2.43
8.7	0.90	2.66	1.13	2.54	1.41	2.46
8.8	0.93	2.69	1.16	2.57	1.43	2.49
8.9	0.96	2.71	1.19	2.60	1.46	2.52
9.0	0.98	2.74	1.22	2.62	1.49	2.54
9.1	1.01	2.77	1.24	2.65	1.52	2.57
9.2	1.04	2.80	1.27	2.68	1.54	2.60
9.3	1.07	2.83	1.30	2.71	1.57	2.63
9.4	1.09	2.85	1.33	2.73	1.60	2.66
9.5	1.12	2.88	1.36	2.76	1.63	2.68
9.6	1.15	2.91	1.38	2.79	1.66	2.71
9.7	1.18	2.94	1.41	2.82	1.68	2.74
9.8	1.21	2.96	1.44	2.85	1.71	2.77
9.9	1.23	2.99	1.47	2.87	1.74	2.79
10.0	1.26	3.02	1.49	2.90	1.77	2.82
10.1	1.29	3.05	1.52	2.93	1.79	2.85
10.2	1.32	3.08	1.55	2.96	1.82	2.88
10.3	1.34	3.10	1.58	2.98	1.85	2.91
10.4	1.37	3.13	1.61	3.01	1.88	2.93
10.5	1.40	3.16	1.63	3.04	1.91	2.96
10.6	1.43	3.19	1.66	3.07	1.93	2.99
10.7	1.46	3.21	1.69	3.10	1.96	3.02
10.8	1.48	3.24	1.72	3.12	1.99	3.04
10.9	1.51	3.27	1.74	3.15	2.02	3.07
11.0	1.54	3.30	1.77	3.18	2.04	3.10
11.1	1.57	3.33	1.80	3.21	2.07	3.13
11.2	1.59	3.35	1.83	3.23	2.10	3.16
11.3	1.62	3.38	1.86	3.26	2.13	3.18
11.4	1.65	3.41	1.88	3.29	2.16	3.21
11.5	1.68	3.44	1.91	3.32	2.18	3.24
11.6	1.71	3.46	1.94	3.35	2.21	3.27
11.7	1.73	3.49	1.97	3.37	2.24	3.29
11.8	1.76	3.52	1.99	3.40	2.27	3.32
11.9	1.79	3.55	2.02	3.43	2.29	3.35
12.0	1.82	3.58	2.05	3.46	2.32	3.38

TABLE 8
Concrete grade C35

$$d'/d = 0.15$$

Redist.	≤ 10%		20%		30%	
$\dfrac{M}{bd^2}$	p'	p	p'	p	p'	p
3.7					0.24	1.07
3.8					0.24	1.10
3.9					0.24	1.13
4.0			Singly		0.24	1.16
4.1			Reinforced		0.24	1.19
4.2	Singly				0.24	1.22
4.3					0.24	1.25
4.4	Reinforced				0.26	1.28
4.5					0.29	1.31
4.6					0.32	1.34
4.7			0.24	1.43	0.36	1.37
4.8			0.24	1.46	0.39	1.40
4.9			0.24	1.49	0.42	1.43
5.0			0.24	1.52	0.46	1.46
5.1			0.24	1.55	0.49	1.48
5.2			0.24	1.58	0.52	1.51
5.3			0.24	1.61	0.56	1.54
5.4			0.24	1.64	0.59	1.57
5.5	0.24	1.77	0.26	1.67	0.63	1.60
5.6	0.24	1.80	0.29	1.70	0.66	1.63
5.7	0.24	1.83	0.32	1.72	0.69	1.66
5.8	0.24	1.86	0.35	1.75	0.73	1.69
5.9	0.24	1.89	0.38	1.78	0.76	1.72
6.0	0.24	1.92	0.41	1.81	0.79	1.75
6.1	0.24	1.95	0.44	1.84	0.83	1.78
6.2	0.24	1.98	0.46	1.87	0.86	1.81
6.3	0.25	2.01	0.49	1.90	0.89	1.84
6.4	0.28	2.04	0.52	1.93	0.93	1.87
6.5	0.31	2.06	0.55	1.96	0.96	1.90
6.6	0.34	2.09	0.58	1.99	0.99	1.93
6.7	0.36	2.12	0.61	2.02	1.03	1.96
6.8	0.39	2.15	0.64	2.05	1.06	1.98
6.9	0.42	2.18	0.67	2.08	1.10	2.01
7.0	0.45	2.21	0.70	2.11	1.13	2.04
7.1	0.48	2.24	0.73	2.14	1.16	2.07
7.2	0.51	2.27	0.76	2.17	1.20	2.10
7.3	0.54	2.30	0.79	2.20	1.23	2.13
7.4	0.57	2.37	0.82	2.22	1.26	2.16
7.5	0.60	2.36	0.85	2.25	1.30	2.19
7.6	0.63	2.39	0.88	2.28	1.33	2.22
7.7	0.66	2.42	0.91	2.31	1.36	2.25
7.8	0.69	2.45	0.94	2.34	1.40	2.28
7.9	0.72	2.48	0.96	2.37	1.43	2.31

TABLE 8 (Continued)
Concrete grade C35

$$d'/d = 0.15$$

Redist.	$\leqslant 10\%$		20%		30%	
$\dfrac{M}{bd^2}$	p'	p	p'	p	p'	p
8.0	0.75	2.51	0.99	2.40	1.47	2.34
8.1	0.78	2.54	1.02	2.43	1.50	2.37
8.2	0.81	2.56	1.05	2.46	1.53	2.40
8.3	0.84	2.59	1.08	2.49	1.57	2.43
8.4	0.86	2.62	1.11	2.52	1.60	2.46
8.5	0.89	2.65	1.14	2.55	1.63	2.48
8.6	0.92	2.68	1.17	2.58	1.67	2.51
8.7	0.95	2.71	1.20	2.61	1.70	2.54
8.8	0.98	2.74	1.23	2.64	1.73	2.57
8.9	1.01	2.77	1.26	2.67	1.77	2.60
9.0	1.04	2.80	1.29	2.70	1.80	2.63
9.1	1.07	2.83	1.32	2.72	1.84	2.66
9.2	1.10	2.86	1.35	2.75	1.87	2.69
9.3	1.13	2.89	1.38	2.78	1.90	2.72
9.4	1.16	2.92	1.41	2.81	1.94	2.75
9.5	1.19	2.95	1.44	2.84	1.97	2.78
9.6	1.22	2.98	1.46	2.87	2.00	2.81
9.7	1.25	3.01	1.49	2.90	2.04	2.84
9.8	1.28	3.04	1.52	2.93	2.07	2.87
9.9	1.31	3.06	1.55	2.96	2.10	2.90
10.0	1.34	3.09	1.58	2.99	2.14	2.93
10.1	1.36	3.12	1.61	3.02	2.17	2.96
10.2	1.39	3.15	1.64	3.05	2.21	2.98
10.3	1.42	3.18	1.67	3.08	2.24	3.01
10.4	1.45	3.21	1.70	3.11	2.27	3.04
10.5	1.48	3.24	1.73	3.14	2.31	3.07
10.6	1.51	3.27	1.76	3.17	2.34	3.10
10.7	1.54	3.30	1.79	3.20	2.37	3.13
10.8	1.57	3.33	1.82	3.22	2.41	3.16
10.9	1.60	3.36	1.85	3.25	2.44	3.19
11.0	1.63	3.39	1.88	3.28	2.47	3.22
11.1	1.66	3.42	1.91	3.31	2.51	3.25
11.2	1.69	3.45	1.94	3.34	2.54	3.28
11.3	1.72	3.48	1.96	3.37	2.57	3.31
11.4	1.75	3.51	1.99	3.40	2.61	3.34
11.5	1.78	3.54	2.02	3.43	2.64	3.37
11.6	1.81	3.56	2.05	3.46	2.68	3.40
11.7	1.84	3.59	2.08	3.49	2.71	3.43
11.8	1.86	3.62	2.11	3.52	2.74	3.46
11.9	1.89	3.65	2.14	3.55	2.78	3.48
12.0	1.92	3.68	2.17	3.58	2.81	3.51

TABLE 9
Concrete grade C35

$$d'/d = 0.20$$

Redist. $\dfrac{M}{bd^2}$	≤ 10%		20%		30%	
	p'	p	p'	p	p'	p
3.7					0.24	1.07
3.8					0.24	1.11
3.9					0.24	1.14
4.0			Singly		0.24	1.17
4.1			Reinforced		0.25	1.20
4.2	Singly				0.30	1.23
4.3					0.35	1.26
4.4	Reinforced				0.41	1.29
4.5					0.46	1.32
4.6					0.52	1.36
4.7			0.24	1.43	0.57	1.39
4.8			0.24	1.46	0.62	1.42
4.9			0.24	1.49	0.68	1.45
5.0			0.24	1.53	0.73	1.48
5.1			0.24	1.56	0.78	1.51
5.2			0.24	1.59	0.84	1.54
5.3			0.24	1.62	0.89	1.57
5.4			0.28	1.65	0.94	1.61
5.5	0.24	1.77	0.31	1.68	1.00	1.64
5.6	0.24	1.80	0.35	1.71	1.05	1.67
5.7	0.24	1.83	0.39	1.74	1.11	1.70
5.8	0.24	1.87	0.42	1.78	1.16	1.73
5.9	0.24	1.90	0.46	1.81	1.21	1.76
6.0	0.24	1.93	0.49	1.84	1.27	1.79
6.1	0.24	1.96	0.53	1.87	1.32	1.82
6.2	0.24	1.99	0.56	1.90	1.37	1.86
6.3	0.26	2.02	0.60	1.93	1.43	1.89
6.4	0.29	2.05	0.64	1.96	1.48	1.92
6.5	0.33	2.08	0.67	1.99	1.53	1.95
6.6	0.36	2.12	0.71	2.03	1.59	1.98
6.7	0.39	2.15	0.74	2.06	1.64	2.01
6.8	0.42	2.18	0.78	2.09	1.70	2.04
6.9	0.45	2.21	0.81	2.12	1.75	2.07
7.0	0.48	2.24	0.85	2.15	1.80	2.11
7.1	0.51	2.27	0.89	2.18	1.86	2.14
7.2	0.54	2.30	0.92	2.21	1.91	2.17
7.3	0.58	2.33	0.96	2.24	1.96	2.20
7.4	0.61	2.37	0.99	2.28	2.02	2.23
7.5	0.64	2.40	1.03	2.31	2.07	2.26
7.6	0.67	2.43	1.06	2.34	2.12	2.29
7.7	0.70	2.46	1.10	2.37	2.18	2.32
7.8	0.73	2.49	1.14	2.40	2.23	2.36
7.9	0.76	2.52	1.17	2.43	2.29	2.39

TABLE 9 (Continued)
Concrete grade C35

$$d'/d = 0.20$$

Redist.	≤ 10%		20%		30%	
$\dfrac{M}{bd^2}$	p'	p	p'	p	p'	p
8.0	0.79	2.55	1.21	2.46	2.34	2.42
8.1	0.83	2.58	1.24	2.49	2.39	2.45
8.2	0.86	2.62	1.28	2.53	2.45	2.48
8.3	0.89	2.65	1.31	2.56	2.50	2.51
8.4	0.92	2.68	1.35	2.59	2.55	2.54
8.5	0.95	2.71	1.39	2.62	2.61	2.57
8.6	0.98	2.74	1.42	2.65	2.66	2.61
8.7	1.01	2.77	1.46	2.68	2.71	2.64
8.8	1.04	2.80	1.49	2.71	2.77	2.67
8.9	1.08	2.83	1.53	2.74	2.82	2.70
9.0	1.11	2.97	1.56	2.78	2.88	2.73
9.1	1.14	2.90	1.60	2.81	2.93	2.76
9.2	1.17	2.93	1.64	2.84	2.98	2.79
9.3	1.20	2.96	1.67	2.87	3.04	2.82
9.4	1.23	2.99	1.71	2.90	3.09	2.86
9.5	1.26	3.02	1.74	2.93	3.14	2.89
9.6	1.29	3.05	1.78	2.96	3.20	2.92
9.7	1.33	3.08	1.81	2.99	3.25	2.95
9.8	1.36	3.12	1.85	3.03	3.30	2.98
9.9	1.39	3.15	1.89	3.06	3.36	3.01
10.0	1.42	3.18	1.92	3.09	3.41	3.04
10.1	1.45	3.21	1.96	3.12	3.47	3.07
10.2	1.48	3.24	1.99	3.15	3.52	3.11
10.3	1.51	3.27	2.03	3.18	3.57	3.14
10.4	1.54	3.30	2.06	3.21	3.63	3.17
10.5	1.58	3.33	2.10	3.24	3.68	3.20
10.6	1.61	3.37	2.14	3.28	3.73	3.23
10.7	1.64	3.40	2.17	3.31	3.79	3.26
10.8	1.67	3.43	2.21	3.34	3.84	3.29
10.9	1.70	3.46	2.24	3.37	3.89	3.32
11.0	1.73	3.49	2.28	3.40	3.95	3.36
11.1	1.76	3.52	2.31	3.43	4.00	3.39
11.2	1.79	3.55	2.35	3.46	4.06	3.42
11.3	1.83	3.58	2.39	3.49	4.11	3.45
11.4	1.86	3.62	2.42	3.53	4.16	3.48
11.5	1.89	3.65	2.46	3.56	4.22	3.51
11.6	1.92	3.68	2.49	3.59	4.27	3.54
11.7	1.95	3.71	2.53	3.62	4.32	3.57
11.8	1.98	3.74	2.56	3.65	4.38	3.61
11.9	2.01	3.77	2.60	3.68	4.43	3.64
12.0	2.04	3.80	2.64	3.71	4.48	3.67

TABLE 10
Concrete grade C40

$$d'/d = 0.10$$

$\dfrac{M}{bd^2}$	$\leqslant 10\%$		20%		30%	
	p'	p	p'	p	p'	p
4.2					0.24	1.22
4.3					0.24	1.24
4.4					0.24	1.27
4.5			Singly		0.24	1.30
4.6					0.24	1.33
4.7			Reinforced		0.24	1.36
4.8					0.24	1.38
4.9					0.24	1.41
5.0	Singly				0.24	1.44
5.1					0.26	1.47
5.2	Reinforced				0.29	1.49
5.3			0.24	1.61	0.32	1.52
5.4			0.24	1.64	0.34	1.55
5.5			0.24	1.67	0.37	1.58
5.6			0.24	1.70	0.40	1.61
5.7			0.24	1.72	0.43	1.63
5.8			0.24	1.75	0.46	1.66
5.9			0.24	1.78	0.48	1.69
6.0			0.24	1.81	0.51	1.72
6.1			0.24	1.84	0.54	1.74
6.2			0.26	1.86	0.57	1.77
6.3	0.24	2.03	0.28	1.89	0.59	1.80
6.4	0.24	2.05	0.31	1.92	0.62	1.83
6.5	0.24	2.08	0.34	1.95	0.65	1.86
6.6	0.24	2.11	0.37	1.97	0.68	1.88
6.7	0.24	2.14	0.39	2.00	0.71	1.91
6.8	0.24	2.17	0.42	2.03	0.73	1.94
6.9	0.24	2.19	0.45	2.06	0.76	1.97
7.0	0.24	2.22	0.48	2.09	0.79	1.99
7.1	0.24	2.25	0.51	2.11	0.82	2.02
7.2	0.27	2.28	0.53	2.14	0.84	2.05
7.3	0.29	2.30	0.56	2.17	0.87	2.08
7.4	0.32	2.33	0.59	2.20	0.90	2.11
7.5	0.35	2.36	0.62	2.22	0.93	2.13
7.6	0.38	2.39	0.64	2.25	0.96	2.16
7.7	0.41	2.42	0.67	2.28	0.98	2.19
7.8	0.43	2.44	0.70	2.31	1.01	2.22
7.9	0.46	2.47	0.73	2.34	1.04	2.24
8.0	0.49	2.50	0.76	2.36	1.07	2.27
8.1	0.52	2.53	0.78	2.39	1.09	2.30
8.2	0.54	2.55	0.81	2.42	1.12	2.33
8.3	0.57	2.58	0.84	2.45	1.15	2.36
8.4	0.60	2.61	0.87	2.47	1.18	2.38

33

TABLE 10 (Continued)
Concrete grade C40

$d'/d = 0.10$

Redist.	≤ 10%		20%		30%	
$\dfrac{M}{bd^2}$	p'	p	p'	p	p'	p
8.5	0.63	2.64	0.89	2.50	1.21	2.41
8.6	0.66	2.67	0.92	2.53	1.23	2.44
8.7	0.68	2.69	0.95	2.56	1.26	2.47
8.8	0.71	2.72	0.98	2.59	1.29	2.49
8.9	0.74	2.75	1.01	2.61	1.32	2.52
9.0	0.77	2.78	1.03	2.64	1.34	2.55
9.1	0.79	2.80	1.06	2.67	1.37	2.58
9.2	0.82	2.83	1.09	2.70	1.40	2.61
9.3	0.85	2.86	1.12	2.72	1.43	2.63
9.4	0.88	2.89	1.14	2.75	1.46	2.66
9.5	0.91	2.92	1.17	2.78	1.48	2.69
9.6	0.93	2.94	1.20	2.81	1.51	2.72
9.7	0.96	2.97	1.23	2.84	1.54	2.74
9.8	0.99	3.00	1.26	2.86	1.57	2.77
9.9	1.02	3.03	1.28	2.89	1.59	2.80
10.0	1.04	3.05	1.31	2.92	1.62	2.83
10.1	1.07	3.08	1.34	2.95	1.65	2.86
10.2	1.10	3.11	1.37	2.97	1.68	2.88
10.3	1.13	3.14	1.39	3.00	1.71	2.91
10.4	1.16	3.17	1.42	3.03	1.73	2.94
10.5	1.18	3.19	1.45	3.06	1.76	2.97
10.6	1.21	3.22	1.48	3.09	1.79	2.99
10.7	1.24	3.15	1.51	3.11	1.82	3.02
10.8	1.27	3.28	1.53	3.14	1.84	3.05
10.9	1.29	3.30	1.56	3.17	1.87	3.08
11.0	1.32	3.33	1.59	3.20	1.90	3.11
11.1	1.35	3.36	1.62	3.22	1.93	3.13
11.2	1.38	3.39	1.64	3.25	1.96	3.16
11.3	1.41	3.42	1.67	3.28	1.98	3.19
11.4	1.43	3.44	1.70	3.31	2.01	3.22
11.5	1.46	3.47	1.73	3.34	2.04	3.24
11.6	1.49	3.50	1.76	3.36	2.07	3.27
11.7	1.52	3.53	1.78	3.39	2.09	3.30
11.8	1.54	3.55	1.81	3.42	2.12	3.33
11.9	1.57	3.58	1.84	3.45	2.15	3.36

TABLE 11
Concrete grade C40

$$d'/d = 0.15$$

Redist. $\dfrac{M}{bd^2}$	≤ 10%		20%		30%	
	p'	p	p'	p	p'	p
4.2					0.24	1.22
4.3					0.24	1.25
4.4					0.24	1.28
4.5			Singly		0.24	1.31
4.6					0.24	1.34
4.7			Reinforced		0.24	1.36
4.8					0.24	1.39
4.9					0.25	1.42
5.0	Singly				0.28	1.45
5.1					0.32	1.48
5.2	Reinforced				0.35	1.51
5.3			0.24	1.61	0.38	1.54
5.4			0.24	1.64	0.42	1.57
5.5			0.24	1.67	0.45	1.60
5.6			0.24	1.70	0.48	1.63
5.7			0.24	1.73	0.52	1.66
5.8			0.24	1.76	0.55	1.69
5.9			0.24	1.79	0.58	1.72
6.0			0.24	1.82	0.62	1.75
6.1			0.24	1.85	0.65	1.78
6.2			0.27	1.88	0.69	1.81
6.3	0.24	2.03	0.30	1.91	0.72	1.84
6.4	0.24	2.06	0.33	1.94	0.75	1.86
6.5	0.24	2.09	0.36	1.97	0.79	1.89
6.6	0.24	2.12	0.39	2.00	0.82	1.92
6.7	0.24	2.15	0.42	2.03	0.85	1.95
6.8	0.24	2.17	0.45	2.06	0.89	1.98
6.9	0.24	2.20	0.48	2.08	0.92	2.01
7.0	0.24	2.23	0.51	2.11	0.95	2.04
7.1	0.25	2.26	0.54	2.14	0.99	2.07
7.2	0.28	2.29	0.56	2.17	1.02	2.10
7.3	0.31	2.32	0.59	2.20	1.06	2.13
7.4	0.34	2.35	0.62	2.23	1.09	2.16
7.5	0.37	2.38	0.65	2.26	1.12	2.19
7.6	0.40	2.41	0.68	2.29	1.16	2.22
7.7	0.43	2.44	0.71	2.32	1.19	2.25
7.8	0.46	2.47	0.74	2.35	1.22	2.28
7.9	0.49	2.50	0.77	2.38	1.26	2.31
8.0	0.52	2.53	0.80	2.41	1.29	2.34
8.1	0.55	2.56	0.83	2.44	1.32	2.36
8.2	0.58	2.59	0.86	2.47	1.36	2.39
8.3	0.61	2.62	0.89	2.50	1.39	2.42
8.4	0.64	2.65	0.92	2.53	1.43	2.45

TABLE 11 (Continued)
Concrete grade C40

$d'/d = 0.15$

Redist.	≤ 10%		20%		30%	
$\dfrac{M}{bd^2}$	p'	p	p'	p	p'	p
8.5	0.66	2.67	0.95	2.56	1.46	2.48
8.6	0.69	2.70	0.98	2.58	1.49	2.51
8.7	0.72	2.73	1.01	2.61	1.53	2.54
8.8	0.75	2.76	1.04	2.64	1.56	2.57
8.9	0.78	2.79	1.06	2.67	1.59	2.60
9.0	0.81	2.82	1.09	2.70	1.63	2.63
9.1	0.84	2.85	1.12	2.73	1.66	2.66
9.2	0.87	2.88	1.15	2.76	1.69	2.69
9.3	0.90	2.91	1.18	2.79	1.73	2.72
9.4	0.93	2.94	1.21	2.82	1.76	2.75
9.5	0.96	2.97	1.24	2.85	1.79	2.78
9.6	0.99	3.00	1.27	2.88	1.83	2.81
9.7	1.02	3.03	1.30	2.91	1.86	2.84
9.8	1.05	3.06	1.33	2.94	1.90	2.86
9.9	1.08	3.09	1.36	2.97	1.93	2.89
10.0	1.11	3.12	1.39	3.00	1.96	2.92
10.1	1.14	3.15	1.42	3.03	2.00	2.95
10.2	1.16	3.17	1.45	3.06	2.03	2.98
10.3	1.19	3.20	1.48	3.08	2.06	3.01
10.4	1.22	3.23	1.51	3.11	2.10	3.04
10.5	1.25	3.26	1.54	3.14	2.13	3.07
10.6	1.28	3.29	1.56	3.17	2.16	3.10
10.7	1.31	3.32	1.59	3.20	2.20	3.13
10.8	1.34	3.35	1.62	3.23	2.23	3.16
10.9	1.37	3.38	1.65	3.26	2.27	3.19
11.0	1.40	3.41	1.68	3.29	2.30	3.22
11.1	1.43	3.44	1.71	3.32	2.33	3.25
11.2	1.46	3.47	1.74	3.35	2.37	3.28
11.3	1.49	3.50	1.77	3.38	2.40	3.31
11.4	1.52	3.53	1.80	3.41	2.43	3.34
11.5	1.55	3.56	1.83	3.44	2.47	3.36
11.6	1.58	3.59	1.86	3.47	2.50	3.39
11.7	1.61	3.62	1.89	3.50	2.53	3.42
11.8	1.64	3.64	1.92	3.53	2.57	3.45
11.9	1.66	3.67	1.95	3.56	2.60	3.48

TABLE 12
Concrete grade C40

$$d'/d = 0.20$$

$\dfrac{M}{bd^2}$	Redist. $\leqslant 10\%$		Redist. 20%		Redist. 30%	
	p'	p	p'	p	p'	p
4.2					0.24	1.22
4.3					0.24	1.25
4.4					0.24	1.28
4.5			Singly		0.24	1.31
4.6					0.24	1.34
4.7			Reinforced		0.29	1.37
4.8					0.34	1.41
4.9					0.40	1.44
5.0	Singly				0.45	1.47
5.1					0.50	1.50
5.2	Reinforced				0.56	1.53
5.3			0.24	1.61	0.61	1.56
5.4			0.24	1.65	0.67	1.59
5.5			0.24	1.68	0.72	1.62
5.6			0.24	1.71	0.77	1.66
5.7			0.24	1.74	0.83	1.69
5.8			0.24	1.77	0.88	1.72
5.9			0.24	1.80	0.93	1.75
6.0			0.26	1.83	0.99	1.78
6.1			0.29	1.86	1.04	1.81
6.2			0.33	1.90	1.09	1.84
6.3	0.24	2.03	0.36	1.93	1.15	1.87
6.4	0.24	2.06	0.40	1.96	1.20	1.91
6.5	0.24	2.09	0.44	1.99	1.26	1.94
6.6	0.24	2.12	0.47	2.02	1.31	1.97
6.7	0.24	2.15	0.51	2.05	1.36	2.00
6.8	0.24	2.19	0.54	2.08	1.42	2.03
6.9	0.24	2.22	0.58	2.11	1.47	2.06
7.0	0.24	2.25	0.61	2.15	1.52	2.09
7.1	0.27	2.28	0.65	2.18	1.58	2.12
7.2	0.30	2.31	0.69	2.21	1.63	2.16
7.3	0.33	2.34	0.72	2.24	1.69	2.19
7.4	0.36	2.37	0.76	2.27	1.74	2.22
7.5	0.39	2.40	0.79	2.30	1.79	2.25
7.6	0.43	2.44	0.83	2.33	1.85	2.28
7.7	0.46	2.47	0.86	2.36	1.90	2.31
7.8	0.49	2.50	0.90	2.40	1.95	2.34
7.9	0.52	2.53	0.94	2.43	2.01	2.37
8.0	0.55	2.56	0.97	2.46	2.06	2.41
8.1	0.58	2.59	1.01	2.49	2.11	2.44
8.2	0.61	2.62	1.04	2.52	2.17	2.47
8.3	0.64	2.65	1.08	2.55	2.22	2.50
8.4	0.68	2.69	1.11	2.58	2.27	2.53

TABLE 12 (Continued)
Concrete grade C40

$$d'/d = 0.20$$

Redist.	≤ 10%		20%		30%	
$\dfrac{M}{bd^2}$	p'	p	p'	p	p'	p
8.5	0.71	2.72	1.15	2.61	2.33	2.56
8.6	0.74	2.75	1.19	2.65	2.38	2.59
8.7	0.77	2.78	1.22	2.68	2.44	2.62
8.8	0.80	2.81	1.26	2.71	2.49	2.66
8.9	0.83	2.84	1.29	2.74	2.54	2.69
9.0	0.86	2.87	1.33	2.77	2.60	2.72
9.1	0.89	2.90	1.36	2.80	2.65	2.75
9.2	0.93	2.94	1.40	2.83	2.70	2.78
9.3	0.96	2.97	1.44	2.86	2.76	2.81
9.4	0.99	3.00	1.47	2.90	2.81	2.84
9.5	1.02	3.03	1.51	2.93	2.86	2.87
9.6	1.05	3.06	1.54	2.96	2.92	2.91
9.7	1.08	3.09	1.58	2.99	2.97	2.94
9.8	1.11	3.12	1.61	3.02	3.03	2.97
9.9	1.14	3.15	1.65	3.05	3.08	3.00
10.0	1.18	3.19	1.69	3.08	3.13	3.03
10.1	1.21	3.22	1.72	3.11	3.19	3.06
10.2	1.24	3.25	1.76	3.15	3.24	3.09
10.3	1.27	3.28	1.79	3.18	3.29	3.12
10.4	1.30	3.31	1.83	3.21	3.35	3.16
10.5	1.33	3.34	1.86	3.24	3.40	3.19
10.6	1.36	3.37	1.90	3.27	3.45	3.22
10.7	1.39	3.40	1.94	3.30	3.51	3.25
10.8	1.43	3.44	1.97	3.33	3.56	3.28
10.9	1.46	3.47	2.01	3.36	3.62	3.31
11.0	1.49	3.50	2.04	3.40	3.67	3.34
11.1	1.52	3.53	2.08	3.43	3.72	3.37
11.2	1.55	3.56	2.11	3.46	3.78	3.41
11.3	1.58	3.59	2.15	3.49	3.83	3.44
11.4	1.61	3.62	2.19	3.52	3.88	3.47
11.5	1.64	3.65	2.22	3.55	3.94	3.50
11.6	1.68	3.69	2.26	3.58	3.99	3.53
11.7	1.71	3.72	2.29	3.61	4.05	3.56
11.8	1.74	3.75	2.33	3.65	4.10	3.59
11.9	1.77	3.78	2.36	3.68	4.15	3.62

9 780863 100291